———THE *SCIENTIFIC THEORIES*———

,,,

Nathan Coppedge

ALL CONTENT

© 2014, '15, '16, '18
Nathan Coppedge

THE SCIENTIFIC THEORIES

Nathan Coppedge

Nathan Coppedge

,,,

———THE *SCIENTIFIC THEORIES*———

SCIENTIFIC THEORIES

INTRODUCTION ——————p. 7
RAZOR OF DESTRUCTION ———*p. 11*

GENERAL ————————---*p. 15*

SPECIFIC AREAS
COMPLEXITY ——————-p. 39
ON THE BIG BANG ————p. 41
ON EINSTEIN-ROSEN BRIDGES—. P. 45
ON MULTIVERSES ————p. 49
ON SPACE-TIME ——————-p. 55
PHYSICS AND ENGINEERING ——p. 59
THEORY OR EVERYTHING ———p. 71
ENERGY AND NO ENERGY ———p. 73
MATHEMATICAL CONJECTURES ——p. 77
ON SOCIAL-SCIENCE ——————p. 100

ON PSYCHOLOGY ——————p. 102
ON SOULS ————————p. 109
ON NATURE ———————p. 111
ON AGING AND LONGEVITY———p. 113
ON CONSCIOUSNESS —————-p. 115
PSYCHIC THEORIES—————p. 119

BIOGRAPHY ———————-P. 124

Nathan Coppedge

INTRODUCTION

So far I am not a well-known scientist. Nonetheless, I have been able to attract occasional interest to my books on scientific topics.

The key reason for this is my long-running website on perpetual motion machines.

Perpetual motion is a theme not only related to physical mechanics, but also engineering, and if we are lucky, innovation and clean energy.

My interest in areas such as philosophy, and art adds an additional qualification to that of perpetual motion: that I could be a connoisseur of literary science.

Indeed, my contributions to philosophy border on the scientific. They often involve such ideas as set theory, complexity theory, and the evolution of the universe.

In this text I simply portray the best ideas I have had about science.

You may be surprised how scientific I am.

Nathan Coppedge

,,,

<u>8</u>

—THE *SCIENTIFIC THEORIES*—

MAIN TEXT

Nathan Coppedge

,,,

THE RAZOR OF DESTRUCTION

In formal logic, basically, falsifiability tends to hold because a theory might be false under absurd conditions. It is almost 100% true that a theory could NOT be false under absurd conditions, so a theory will remain largely falsifiable under those absurd conditions, if we assume most conditions are falsifiable under the specific theory.

Even though, MOST theories are falsifiable under ANY theory.

The obvious thing is that something falsifiable could later become unfalsifiable, however this is not true in science. However, science does not hold the ultimate claim: reality does.

Falsifiability seems to claim that the universal cannot be known*, although it can become more and more likely that new truths require new approaches (one or plural).

There are also other concerns rather than just the macro picture. It is possible the data is wrong about what is true (truly represented) regardless of whether it is falsifiable— thus, falsifiability is not the first question to ask. Accuracy comes up before falsifiability.

Also within the details, there may be premise logics which make more assumptions than science and draw sound universal conclusions. After all, based on conclusions alone they do or do not have accurate conclusions in some degree, and so, logically the conclusions are accurate or not in some degree, and if any conclusions were absolutely true science would be claiming universals were not falsifiable, if falsifiability depends on a theory.

In some cases truth (logical or empirical) may come first even if truth is difficult to guess — in which case it can be a fallacy to assume what is truly accurate is the easiest to guess.

This last point is more certain if there is any type of logic which can draw the same conclusions as science.

We should not assume empirical and logical truth are always different, but if there is no resemblance at the very root of it, the conclusion is that science is either logical or empirical, but not both.

So, if evidence cannot be reproduced logically, this suggests it is not mathematical.

And, if science IS mathematical, then this suggests universals would not be falsifiable.

But if we assume universals are falsifiable, we may be forced to reject mathematics.

So, how far would we go to accept the scientific method? Would we go so far as to defend logic? Or, would we go so far as to reject mathematics by rejecting universals?

It looks as if science is purely logical and mathematical, but if so we should reject the scientific method.

* (Notation: Even if knowing is different from theory, knowledge may re-

quire theories to have a language).

- If science cannot be guessed by logic even in retrospect, it is not provably logical.

- It is then no wonder that science cannot guess universals, since if it is not logical, it is not logical.

- Then science should be guessed by logic unless it is not logical. If it is not logical it is incommensurate with mathematics.

- If science IS logical, goodbye to universal common sense, because what is truly accurate may not be the easiest to guess.

———THE *SCIENTIFIC THEORIES*———

GENERAL THEORIES*
This section is chronological.

What social experiments are interesting to observe: Whatever's bad is interesting to observe. But if you know good and evil you don't like bad.

Alternate explanation of dark energy: If the force is associated with gravity, 0.5 mass * distance involving the solution to the three-body problem may explain it, e.g. if a third body pulls a 2nd body over a first body, the property may be similar to perpetual motion.

Testosterone is not the essence of success: it is possible not to be rich or not to have strong muscles with fair amounts of testosterone.

Zheng Guo couldn't shake the thought that Chinese apes looked a bit Chinese, but he wouldn't admit it. Darwin admitted it.

It is expected there is a slight flare-up when you apply [miracle cure]. Because the earlier [medication] is designed to make it look like there is no cure.

Coherent A.I. Research: Humans are normal to be self-aware, and godlike to think. Humans think it's easy to think, but can't do some basic things. A.I. is normal to think, and godlike to be self-aware. A.I. is aware it's hard to be aware, but it can do some pretty advanced things.

Salient research: I feel like a bu**head when the bridge my nose is indented.

Are rats eating plastic?
Two-component families are normal, unless someone knows how to reproduce. How someone feels always depends on the quality of their treatment more so than the quality of the food they consume.
*Here treatment means mentality, metaphysics, and meaning to the exclusion of drugs.

Relaxation is always a response to stress.

Madness always has a better theory to a wider point of view, with sufficient imagination, until a new form of reason is found. Madness is meta-sanity, and reason is meta-madness.

Someone will invent a material that is immaterial. Things will be turned on their head.

Where psychology does not have a bond, there must be some other kind of common bond.

THE *SCIENTIFIC THEORIES*

Soft drink fizz sometimes contributes to skull-expansion, if not brain expansion.

If we had a truly relaxing color, we could feel much more secure about productivity, because the night hours could be replaced by intermittent spells of relaxation.

People can understand aggressive behavior because it's a more common problem than some others. This creates a false sense of dependence on anger.

Caffeine causes burps.

Those who suffer from paper mites don't believe they can do anything with paper. This assessment has proven true time and time again.

If similar equipment is used to detect ultra-violet waves as to detect radio waves, perhaps there are other types of waves that require different equipment. There is a hint that some of this has to do with classifications of matter, which continges upon available perceptions. Some forms of waves that are now considered meaningless may someday be considered as being equivalent to radio waves. For example, the waves of the ocean.

Conjecture: People who borrow clothes

gain weight in the long term. Or, perhaps most people gain weight. But it is possible this has a conditionality. In other words, the borrowing of clothes and the gaining of weight may be equally justified, and even for related reasons, under some rhetorical stress.

Eating certain very healthy foods reinforces the tolerance for drugs such as sugar and medication. This is often an understated fact.

When detecting people in wreckage, people are more likely to notice someone who is subject to heart attacks, because their heart beats louder. Athletes who suffer from heart attacks survive wreckage with the greatest frequency, unless they have a heart attack while they're being saved. Exceptional rules could emerge from this type of analysis.

'Paradigmatic replacement sets a standard for social science.' Rules like this are themselves paradigmatic.

Optimalism is anethema for parasites in dimensional worlds. I have found that conceptually, if the proposed dimensional being takes variables, and the variables tend to improve automatically, there would be no room for parasites that do not

THE *SCIENTIFIC THEORIES*

beneficially co-adapt with the organism.

Force tends to be a vector of displacement. Pure energy displaces vectors, or true displacement is a displacement of force. While based on classical physics, these types of principles derived from a deductive method may prove radical in determining the true properties and functions of materials in the universe, and hence, beyond.

The criterion of genuine science is that it instructs nature. But how could this take place? True instruction is a relation to nature, or involves a relativization of criteria. In this sense, instruction is an application more than it is a science. For, nature that is scientific must be understood before it is demonstrated, or science is very similar to nature, because only in nature alienated from science are lessons possible which do not involve understanding. If it is understood that the lesson follows from the understanding regardless of when the lesson occurred, it is more realistic to see science as an application of nature---or some other idea, rather than nature as a formulation of science---since the two formulations are technically opposites, perhaps more so than the terms 'science' and 'nature' themselves.

Game states are derived from a figurization of the object-condition. If the object-condition cannot be figurized, little emerges as a game, and thus, little emerges for the prospective science of games. Apparently, the biggest unknown for the type of science that concerns games, is the territory which cannot be figurized. But the biggest concern for science in general if games are important, is whether science concerns games, not whether games can be figurized. Thus there is a double-bind. It would be better to approach the question of figurization first, before the question of games, if it is uncertain whether games are important. That much is certain. For an erroneous conclusion about games in general is more egregious in the long term than an accurate conclusion about whether games could be theorized, or how priorly un-figurizable games can now be considered, through the science of the un-figurable. Under the theory that only one thing is the 'un-figurable' ---a theory I take to be true conceptually----many things ought to be figurable, and no relativistic argument in defense of the unknown should be substantiated. The only major argument against this is the argument that 'only one thing *is* figurable' which can be taken to mean the valid things within the context of any present application. In this

THE *SCIENTIFIC THEORIES*

sense, there is a value for generalities, but it is also clear that the figurable and the un-figurable are equal identities in the equation of determining an application. Even if all this means is negative space (*wu wei*), there is an avenue here for theories which work from the unknown, and extrapolate towards the known. What is not known in a sense always means something which is desired to be known, even if it can only be theorized. In this sense, what is unknown is not merely theory, but instead, gradually highly empirical. Otherwise, science is sacrificing imagination for the sake of the unknown, a theme which can easily be discredited, since conventionally the unknown means very little.

It may be easier to breathe when it is the temperature for yogurt cultures.

One theory is that complexity ages. However, if it ages, it must grow less materially complex as it breaks down. There is then no choice but to believe that complexity does not age, or that complexity grows immaterial. Perhaps complexity is a dimension which exists independent of entropy.

What might be impossible, it is argued, is always an application of impossibility. By this argument, nothing is actually impossible, since the application is a formality. Without the formality, things cease to be impossible. They may equally cease to be anything at all.

It must be remembered that context-fields used in statistics or categorical deduction are symbols, and symbols can be manipulated.

Truth (heuristic, rubric) is the divide between pleasure and pain.

We are dead, or language is the carapace of society, or (perhaps with magic), the universe itself is a carapace!

Unless it's authentic willpower, there's no point in being pro-active.

Knowledge is shared. So many people have the same clues to cognition.

Structural fire could exist, it's just that scientists had been thinking of it as a gas.

THE *SCIENTIFIC THEORIES*

Many scientific theories advance like this. It could also be called molecular fire. Inconveniences in words often cause vast gaps in research integrity.

The weakness of sugar might be psychological guilty feelings, and the genetically inherited (from honey) weakness that bees get when they are young and drunk on honey. In this sense, it may be a sense of genetically inherited youthfulness, which may destroy muscles, but does not destroy the core concerns of health, except through psychological damages.

Chewing costs energy. Sex costs energy. Thinking costs energy. Almost none of the pleasures are leisurely in the way people imagine them to be. People are duped about leisure.

Addicting matter in a certain way may be like creating matter. Perhaps there are cases that are not cancer.

Clinical research that is not as religious begins to be more creative. This has been observed on the case of constructing Hyper-Cubism and not receiving adequate psychological stimulation. If the experiment were conducted properly, research would in no way inhibit the creative process. Apparently, it would promote it, since the desired subjective effect is positive.

Avoiding self-love forces men to confront their issues, creating an intermediate period of violence for those that have inner conflicts. The ultimate result, however, is spiritually redeeming. Thus, there is a choice between accepting conflict and accepting degeneration, a hard bargain. One may wonder why politicians do not embrace a religious view, if a major component of religion is rejecting masturbatory idealism.

Pain is not a motivator for authenticity. The rejection of this principle has been a foundation for the simplistic Pavlovian view, such as that favored in motivating armies, and perpetuating warfare. If pleasure cannot be guaranteed, then meaning should be provided. If meaning cannot be provided, pain should be prevented. If pain cannot be prevented, pain should be minimized. If pain cannot be minimized, pain should be explained. If pain cannot be explained, however, then life is hell.

This may be a small detail, but it appears that the most genuine variety of perpetual motion inventors are not technically obsessive personalities, in the negative sense of

THE *SCIENTIFIC THEORIES*

the term.

Human brain chemistry: What isn't serious is an encounter with dopamine or other pleasure systems, and it goes downhill from there. As always, one solution is to evolve!

In certain light-up elevator buttons, I have observed that making a gesture towards a second push of the button without pushing the button a second time will improve the probability that the button registers the first push. Although one could say that the gesture at a second push causes the first push to register harder, there is no way to prove this if the only way to determine if the first push is harder is whether the first push causes the light to light up. Therefore, one might as well compare an ordinary single-push and an ordinary double-push with the single push followed by a half-intentional feint. But even in these cases, if the effect is caused by some sort of magic is un-determinable, unless the effect always works for some people and not others. But are those people necessarily people with a hard double-push? And what would that say, if, after all, the case involving a feint doesn't involve a second push? At what point can it be determined that any special effect happened at all? This is one of the only cases I've found where the laws of physics are arbitrated not by principles, but by common sense.

If we could prove determinists were believers in fortuity, we would have a basis for thinking that the motivations of determinists might be purely psychological, rather than rational. Why not rational? Because, if something is psychological, this suggests that someone is getting what they want, rather than what they do not want. And reasonably, in the world in which one does not get what one wants, there is evidence of determinism---a lack of control. But if one gets what one wants, it appears that one has influence --- which is like control --- which is like willpower. Thus, if determinism is purely psychological, even if that is sufficient for determinism, it also serves as evidence of free-will.

On top of the free will theory is the theory that science is largely un-provable.

You will find that if children are told to imitate trolls, the more important the activity is perceived to be, the more intelligent the children will perceive the trolls to be. In fact, no matter what humans behave like, it is that thing they behave like that is perceived to be intelligent. For example, for the Native American Indians of North America, behaving like a coyote or wolf

THE *SCIENTIFIC THEORIES*

was perceived to be highly intelligent, due to the belief in a trickster god named Coyote. If there is an exception to the rule, it is because the human perceives the thing being imitated to be under human control. When it is not under human control, it is seen as possessing great intelligence. Otherwise, it is not a worthy object to imitate.

Imagine a meteor shower: better many little ones than one really big one. At least in the case of little ones, roofs can offer protection.

Consciousness is an outlier. Because if consciousness is not an outlier, then the reality that is expressed is absolutely real, meaning that all the people that are observed are real people. If they are not getting what they want in every case, the acceptance of their absolute reality bring us to the conclusion that their ailments are real, but cannot be explained by their desires. So either consciousness is an outlier, or people always get what they want, or there is no way to explain pain, because pain cannot be explained if it is not what someone wants. If people always get what they want, then there is no way to argue that something has gone wrong. This view denies that anyone experiences excessive suffering. Nor is the view that suffering is unexplained a good answer. A view that accepts unexplained suffering must also accept the problem of evil as though it is unavoidable. So the exception to explaining away suffering, and unexplained suffering is apparently the view that consciousness is an outlier, which is like a Nazi view. Although the Nazi view might accept a wide variety of skin tones, etc., it is not likely that it would accept an unexamined life, except by imposing the idea that some people are more real or more useful than others. Therefore, there is a double-bind between the Nazism of the majority (conformist, ordinary-looking people) and the Nazism of the minority (self-examining, unusual-looking people), that meets its foil with the opposite categories of conformist, unusual-looking people, and self-examining, ordinary-looking people. Therefore, apparently the concept of elitist consciousness is opposed by the popularity of philosophy and unusual appearances, although it is these very things that promise to preserve the concept for the majority. No wonder intellectuals tend to go crazy, since these concepts clearly relate not only to political concepts, but also to self-identity and individual philosophical theories! To be sane, the individual has to align with the majority, which involves explaining pain away, or leaving it unexplained. But the self-examined life leads to the rejection of these views.

THE *SCIENTIFIC THEORIES*

Authentic boredom is more exciting than nothing at all, unless the nothing is itself authentic.

A subtle science: something missing from science. We can predict subtlety by predicting what we don't observe. First, we say that what we observe expresses knowledge. Then, subtlety concerns the knowledge that we don't observe. So subtlety has the opposite properties of the knowledge we observe. Subtlety has the properties of what we do not observe. 'Thank you for filling in my mind' someone said.

Theories of nature: The more perfect humanity becomes, the more it is surrounded by waste-products. Maybe beer, or diamonds, or art supplies, or loud movies, or startling images?

Theories of people: People defined by the relation between the functions of their self and their true self. Therefore, 'desperate' people, 'idealist-maximizers', 'politicos', and 'children-waiting to grow up'.

Theories of technology: People are always discovering the tip of the mountain rising out of the dimensional skirts of data. They are re-acquainted with their former selves, and sometimes think that they have arrived at a new type of identity. New technology is where people appear foolish. Technology is about access to the universal, and the universal lexicon.

Theories of purpose: Purpose is either emergent, like technology, or characteristic, like people, or an outer experience, like nature, or some other formula. Politics is often the third kind. War is the second kind. Philosophy is the first kind. Theories of purpose can create mistaken definitions, much like half-useless technologies.

Psychology of Science: Exceptional Convenience: The observable is explicable from the outside, unless it is self-explanatory from the inside. It is either a case of convenient observation, or convenient reality: in either case it is a case of convenience with science.

Time and heat are variables. Heat is a two-dimensional variable. Perhaps heat-time is an expression of the universal energy. Matter is a theory of truth like heat is a theory of time.

The belief in counteraction is a thesis I widely do not support. Scientists and psychologists seem overly fond of this word, and in my view it creates a kind of false-

––––THE *SCIENTIFIC THEORIES*––––

positive mirror-image. That is easily ignored while reflection connotates illusion. Real theories are more concrete, and even aspire to be more *com*plete.

It must be remembered that if memory is genuine, then stimulation is partly memory.

Concerning universal anthropology: To the dimensional sense of the person, the central motif appears to be satisfaction, or some variation on it. Species that do not develop satisfaction are purely logical, and easily manipulated.

Things like language proliferate, and with them, a wide variety of logics----or other, far more obscure terms. If we are too enmettled to realize this, at least it serves the value of our complexity.

Identity, it strikes me, is to a huge degree an open-ended concept. There is nothing about identity which says that it does not oppose itself, or that it did not adopt some difficult obscurity for its realization. From this type of principle I arrive at the idea that identity is universally exceptional.

If a fixed screen at a closer distance produces a ripple effect when there is motion on the opposite end of some bars, then perhaps the light-wave effect is also produced by a kind of difference----in this case a difference between the infinitely detailed and finite perspective. The infinite perspective is analogous to a screen with infinitely small holes, whereas the bars are just how the light happens to be at a given time---often more than one particle, thus, finite, but within an infinite mesh.

Chocolate stimulates dreams. Chocolate may be the major cause of dreams. Thus, elitism, as shown by rewards, means that dreams are a major difference between the elite and the non-elite. Has political science ever been so biological?

Jews tend to be non-coherent philosophers, because they have been circumcised. They tend to think that systems are about augmentation, rather than 'encircling'. It is a major difference. Hence visions of infinite cities and the influence of God, but a mostly incoherent mathematics.

Those who desire to bask in the sun after drinking milk very often have a worm.

—— THE *SCIENTIFIC THEORIES* ——

There was an article which said that ears developed much similarly to funguses. And another, I believe, that said the phallic part of the body did the same, in response to gradual evolution and the messy contents of the female sex parts. What is then to stop us from thinking in some ways, that an animal is just a more advanced plant? Clearly there is some analogy between ears or the phallus and photosynthesis or brooming (which are both plant processes…). Perhaps this thought is owed to Aristotle… If he had the word 'advanced' he certainly would have thought of this. Perhaps he would have said that the animal is the *telos* of the plant, which is profound enough…

I suspect gravity occurs through magnetic wave-avoidance. This would explain the vertical character of gravity.

Specific ideas, such as relativity and space travel can be radically exponential when interpreted in terms of the effect upon the character of physical bodies and how they interact with the environment.

Methane is like a second kind of air, that can be tolerated by nature (the source), but not by second nature (other people, civilization).

Drinking brandy can seem like eating an apricot, for those that only have eaten apricots, if brandy is perceived as apricot brandy.

Analysis is just as effective as the scatological in stimulating scatology.

By and large, if people were not attracted *a little bit* to things, they would never gain an addiction. Thus, subtle addiction are responsible for un-subtle ones. Part of this is merely the ability to cope with semantic distinctions. And, another part is depressions that some people have, that others take as serious rhetoric, compromising the semantics. For, originally, everything WAS good when it was good, even though it is hard to remember that. But we shouldn't AGREE not to remember. But it is the opposite with addictions. We are supposed to remember the future and forget the bad things. Hopefully some day addictions will be considered delusional.

―――THE *SCIENTIFIC THEORIES*―――

A semantic theory of health holds that the reason that companies don't hold standards is because some people don't hold standards. Thus, poor standards are responsible for poor health.

What may be deduced to be rancorous, either ethically, or through a weird potency, about an addiction, makes its sacrifices ultimately more tolerable, whether these sacrifices take the form of recovery, or long-term abuse… The fundamental habit of the addiction is being accustomed to it. And, it is the uncertainty of the habit that grants serious drugs their most common edge… Uncertainty is to uncommon consumption what certainty is to common consumption!

The human body has a fixed capacity to gain weight, based on the stored energy that has already taken place. This is a way of explaining the relative durability of thinness from one person to another. Also, in a dimensional sense, the earlier one is fed the earlier one compensates for the 'first difficulties' which are the source of the deepest pangs of hunger.

Open question: 'All of what perception could be anything consistent?'

General theory of scientific mistakes: Some part of the intestines may have a weak wall that allows parasites to creep in. The scientist chalks it up to coincidence, while the superstitious person thinks it's a result of a relationship with the parasite. It may be that the superstitious person is right. If we take the superstitious person's point of view, the weak wall of the intestine can be explained by the parasite. But if we take the scientist's view, the weak wall is inexplicable. Therefore, the superstitious person is probably right.

Students should be encouraged to believe that knowledge is highly accessible and often at its core, non-technical. Otherwise gaps appear in the history of knowledge, and an important intuition is lost.

The world does not help you if you have a slightly above average I.Q. If you have a moderately high I.Q. you can solve problems. If you have an average I.Q. the world helps you solve your problems. If you have a high I.Q. you can complain about the problems you don't have. If you

THE *SCIENTIFIC THEORIES*

have a below average I.Q. you can enjoy your problems and still think there's something wrong, or even ignore the idea that you feel good.

Often the theory that everything is stupid becomes interesting at precisely the time that someone loses I.Q. The same theory might be uninteresting at the point that the same person had lower I.Q. before.

According to one doctor, sweating profusely in a cotton shirt is what causes asthma.

It may be that the common cold is caused by bone marrow depletion.

Re: Razor of Destruction and Sub-Razor of Excitement :It is possible all viewpoints which are not excited involve some form of chemistry.

Bile might be produced by pride-tagging.

Nathan Coppedge

,,,

COMPLEXITY

You may have come across this term. Complexity is concept-limited. That means that, although the universe may be infinite substantially or interpretively, actually being infinite in a measurable way tends to involve an infinite definition, an infinite capacity, and an ability to realize dynamic infinity.

Consider this thought experiment. A particular area of space contains a particular category of description. But, it's energy content is limited. Therefore, its potential expressed in energy is not infinite. If it has infinite potential, it has to be some form of efficiency that does not make use of all of the energy.

On the other hand, a finite definition can have infinite content. It is called proportional numbers (some call these 'ordinal numbers' confusingly). It does not always have infinite energy. In fact, it rarely does, unless by definition it is the sum total of some infinite quantity.

Therefore, there are four choices:
1. Finite definition, finite energy.
2. Infinite definition, finite energy.

3. Finite definition, infinite energy.
4. Infinite definition, infinite energy.

Unfortunately, we don't have much capacity to comprehend infinite definitions, so our concepts are likely to be limited to the finite. When we find an efficiency, we are likely to think that it, too, is concept-limited. But this may not be the case.

Thus, we can adopt a functional standard, in which what is infinite is an infinite function, and what has energy is what functions with energy (heat, for instance, or causality). Whenever the capacity is sheerly relative, then there is no way to measure whether a thing is infinite, or whether it has energy, except in relation to other things.

Thus, definitions remain concept-limited, but not necessarily inefficient.

That becomes the essential definition of infinity, for the universe or anything else.

THE *SCIENTIFIC THEORIES*

ON THE BIG BANG

It might be said that we are still at the beginning of the Big Bang, relative to the future, assuming the expanding continues. And, if instead we are not at the beginning, it is likely that we haven't yet found the true beginning, if it actually exists, because we would then live in a longer time frame than we have predicted so far.

So, there is a likelihood that while we have a causal relation with matter, then we are participating with the Big Bang. When there is a causal disconnect, then we have entered a different universe. Either of these may be possible, and a different universe is not necessarily what we mean by returning to the Big Bang, but it is not necessarily what we mean by NOT discovering it, either.

I hold to the firm rule that the meaning we find in what we discover about ourselves is the most determinant factor in the whole causal development. If we don't find meaning, that is the end of the universe, relative to US. But if we find meaning, then there is likely some exception which will grant us some type of universe in the future.

Consider the following thought experiment: Something artificial in a very real virtual reality, with excellent senses is just as real as something extremely real that is extruding within a sort of poor-quality virtual reality. So, unless the difference is cognitive, it appears that the universe is much more real than we are, and is therefore subject to considerable subtlety, or the universe is a function of perception within virtual reality, which scales to the degree of perception relative to the reality of the program we live in.

But if virtual reality is just a passing fad of history, then clearly the universe evolves, and the ancient concept that the universe is eternal probably trumps recent theories that it is mortal. Why not explain these recent theories as the gods of the ancients closing their browser window?

In that case, discovering the Big Bang becomes sort of meaningless, because the real thing was Divine Creation, much more actual than the real Big Bang of our own universe, which barely exists at all. In our universe, it probably just exists for entertainment, if humans are taken as important. But if humans aren't important, there must be a better explanation than the idea that the universe is eternal! Or per-

THE *SCIENTIFIC THEORIES*

haps, there could be further concepts beyond the universe, that explain human aspirations.

Another principle I have is that nothing happens except by imitating something more expensive that does the same thing better. So, by this principle, there is infinite potential, and it is just a freak coincidence that we're stuck in a cheap mortal world that is not composed of ideas.

Nathan Coppedge

,,,

44

ON EINSTEIN-ROSEN BRIDGES

Black Holes and Wormholes: Principles:

*Wormholes have the dimensions of the location in which they are found.
*It may be possible to traverse a wormhole without knowing the number of dimensions that it consists of.
*The more dimensions the wormhole has, the more easily it can be traversed (I think, since more dimensions imply more degrees of freedom).
*Traversing a wormhole may depend on the wormhole's being more dimensional than the black holes near where it is found, in the case of black hole-type wormholes.
*It may be the case that it is easier to create a wormhole without black holes.
*A black hole likely requires a wormhole that has a large number of dimensions, but not significantly varying from the number of dimensions of the black holes.

Time-Travel to the Past Involving Relativity

That depends on whether the space-time around us is in a smooth, continuous loop with the past, or if it is experiencing some type of 'time-flux'.

In the case of time-flux, we might age more quickly, or experience the world passing by very rapidly, while we remain relatively the same age. Whereas, in the case of a smooth, continuous loop, we age at a regular rate, and merely encounter the world around us progressing at whatever the rate of change is between the past version of the universe we are traveling from, and the future version of the universe we are traveling to.

So, you can imagine that with a very fast wormhole, we want to travel very quickly, or the universe we travel to will be much older than the one we're traveling from. On the other hand, with a slow wormhole, we want to travel very quickly, so that we don't age enormously before reaching the destination universe, which may still have a nearly identical age to the present one.

That's the basics, if we assume it's wormholes. But it gets much more complicated if it involves semantic travel, because that would mean that our very thoughts come into play in how we PERCEIVE the present and the past. In other words, memory comes into play, and sometimes the age we experience is only a variable, rather than an actual time-dimension.

THE *SCIENTIFIC THEORIES*

If we could travel contingent to time, we would have a key to immortality. But time-entropy might predict that this is impossible. But if there is no time-entropy, then travel may be used to achieve immortality, at least vis. the current dimension of time.

2021–11–23 It was found there are 25 fundamental ways to change universes:

0. Forget impossibility.

1. Adopt incoherence.

2. Stop growing.

3. Get stopped.

4. To not be naturally mathematical.

5. To stop being rare.

6. To lose your identity.

7. To be unemotional.

8. To be unimportant to humans.

9. To be dysfunctional.

10. To be un-intellectual.

11. To be easy.

12. To be coherent.

13. To be primitive or posthuman.

14. To be unfortunate.

15. To be unclear.

16. To escape time.

17. To be healed.

18. To have effects that disappear.

19. To not communicate.

20. To gain rationality.

21. To go towards the center.

22. To be ugly.

23. To not have a special advantage.

24. To significantly lack opportunity (though this is bad).

———THE *SCIENTIFIC THEORIES*———

ON MULTIVERSES

I discovered recently that certain types of abstractions are believable and yet with only thought as evidence. It seems to vary from person to person what these things are. I think it points towards an as-yet-unrealized reality in higher dimensions. --- Quora 2016

A.

I have several interesting theories on the subject of multiverses:

(1) Fossil Realism: the further we look into the background, the further into history that we go. In other words, the more we look for proof, the more we find permanent structures, and the less we find subtle ones.

(2) Fossil Realism can have the effect of implicating consciousness theories. That is, on the subtle level, the universe is the least descriptive thing. For example, it is possible to construct a thought experiment in which there are two worlds. In the first world, which has the same macro model as the second world, life is not worth liv-

ing. In the second world, life IS worth living. Since the difference is apparently determined by a 'CHOICE', it is more future-determined than past-determined. In other words, it concerns values more than beliefs, and systems of logical prediction (qua meaning) more than observations.

(3) In some ways, all causes are products of the most fundamental causes. But it is hard to determine what these are. For example, one could look into value theory or justice theory, all implicating theories of atemporality.

(4) There also may be a difference between VISIBLE physics, and REAL physics. What if two different microwave backgrounds with the same physical properties according to our measurements, have DIFFERENT FEELINGS? How do we come about to detect what that difference means? We cannot just say, "it is subjective" and call it a day. Apparently, there is a way in which variation is more about similarity than it is about difference. In this sense there is a relative absoluteness to accurate claims, and a quantum arbitrariness to merely theoretical ones. Yet some people find that

relative absoluteness does not describe anything at all. But if it speaks to our subjective experience, we have to say that it concerns the universal, even though we do not know that it is the SAME universal that someone else is referring to.

(5) The project of objective data begins in polar opposition, between (1) what is TRUE TO US, and (2) THE BEST WAY OF PORTRAYING IT. Only as a second step do we reach something which could be called universal, which is the unique character of reality, which is both TRUE TO US, AND ACCURATELY PORTRAYED. There is a radical need for non-arbitrary data, which runs contrary to science.

(6) It is understandable if science runs against mere imagination, but if imagination is one of the faculties which allows us to OBSERVE WHAT SOMETHING IS, then we have to accept that there is some overlap, in all relevant data.

(7) The problem being, we do not know if we are observing different universes. So we have to begin with the arbitrary, proceed through what is true to us and what is the best portrayal, and end up in what is

not arbitrary. My overall impression from all of this is that there will be a greater role for the semantic scientist in the future. For example, consider that someone may think that some or another large chunk of the universe is more important than another. In some sense, this is a DIFFERENT PHYSICS. By sheerly determining what is important, it is as though there is a different map, a different reality. The only reason we have not discovered this, is that we have never had TOO MUCH GENUINELY RELEVANT DATA. We have always had a stark picture, not a real picture, or not a desirable picture. In the ideal reality, multiverses are a semantic point, which takes variables like degree of investigation. Perhaps that is actually, somehow, the final word on the matter for now, whether or not anyone believes me.

B.

I guess what I meant to say is, there's no rule against material semantics.

C.

Another thing to look into would be if bubble-verses are not, in fact, bubble-shaped.

There is a possibility here that multiverses behave through some sort of sucking worm-hole type activity, or even something more complex.

As a corollary to this, perhaps researchers should investigate complexity issues, such as 'how much channeled flux occurs' relative to complexity.

My intuition on this is that there is a lot of participation in multiverse-type activity in terms of a single element, but less evidence of cross-identification.

In the first place, a common element can undergo semantic ambiguation, an active process that results in some changes of properties. But it is not precisely categorical. In the second place, trans-elemental activities require some sort of logical reason for correspondence, and thus, semantic ambiguation is less likely. However, semantic ambiguation COULD occur through some variable medium (emphasis added) such as energy, relativity, or quantum background.

Nathan Coppedge

,,,

54

THE *SCIENTIFIC THEORIES*

ON SPACE-TIME

[How Is Time Affected by the Expansion of Space?]

I don't know if there is any scientific consensus on this.

Around the time of Einstein, the popular thing was to believe that the two are mutually continuous. In other words, time is just another dimensions of space, that is, it is the fourth dimension of space.

More recently, talk about multiverses has complexified the issue. What if, for example, there are multiple time-streams, and interaction between the streams? How then is time 'just one dimension of space'?

My own conclusion is that time is a metaphysical variable. When a second dimension of space-time is genuinely contingent to another dimension of space-time without a necessary time-vector, then immortality and absolute free will are possible.

In lower dimensions, time itself may not be a constant, and existence is experienced intermittently as an 'access process'.

Now, what if further dimensions of time exist which are not yet immortal? In that case, we get complex time, which involves experiencing time like it is a form of space.

In the third or what we now take to be the fourth dimension (but which I take to be a variable applied to the third or third-and-a-half), the experience of time is simplified and one-dimensional, unless we succeed in traveling between dimensions. This form of travel might occur constantly, although it is sometimes invisible since the dimensions are so similar (if the variation is spatial variation, but not physical variation due to isomorphism).

Time travel occurs in the fourth dimension, and can be experienced in the three-and-a-halfth. The ideal sense of the fifth dimension is immortal, although it may still involve perceptual flux. Thus, experience of time in the true sense in the 3rd - 4th dimension involves time-travel, while experience of time in the 5th and higher dimensions can treat time as a form or object.

In any case, time appears to act on the objects that interact with it, either through willful change (in the 5th dimensions), or through voluntary or involuntary travel or simple experiences of the greatness of the

THE *SCIENTIFIC THEORIES*

universe (in the 4th and lower dimensions).

The sense of simple experiences of the greatness of the universe suggests that time is something like traveling space. This explains why it requires travel to act against it.

Therefore, there is a strong relationship between a person or object's ability to self-determine itself, and its ability to manifest and interact with time.

[If the Universe was to end, would time also stop? Or would it continue its infinite sequence?]

One good answer to your question is that the universe always surrounds something alive.

Therefore, before the universe ends, the life within the universe must end.

Thus, unconsciousness precedes the end of the universe by an indefinite amount of time.

The universe probably exists as some form of 'potential' even when it has collapsed or lost heat energy. But, it would look like nothing to us.

Again, we're lost in the concept that life must have meaning of some type before it can signify anything.

This sort of significance, which Frege described as 'denotation before connotation' traps us into thinking that the universe will end, when in fact, even nothingness may have infinite untapped potential--- even if it is just ultra-subtle information.

However, space can collapse and expand, and this results in a different consciousness of what is meant by the current location. In a sense, where we are located can be different even if objective location (if there is one, even relativistically) remains the same.

It is this kind of paradoxical space-time that we have to confront if we are to determine the ultimate nature of collapsion and expansion.

THE *SCIENTIFIC THEORIES*

ON PHYSICS AND ENGINEERING

Quantum Invisibility Waves:

*Possible Breakthrough
for Quantum Invisibility*

Developments:

- Quantum invisibility waves: simply sees large 'planets' as hidden by quantum invisibility: a black hole would actually be an extreme gravity planet seen from great distance, with light failing to bend through it

It is possible *failed observation of macro-scale phenomena does NOT imply knowing the probability, to the degree that something possesses mass.*

This follows from the fact that the properties of sub-microscopic phenomena are especially probabilistic, and therefore reveal an aspect of themselves due to polarization. At macro scales, the polarization is less evident, and the object becomes less visible in proportion to mass.

Another way to put it is:

The specific probability of macro-scale objects may be a previously unrecognized unknown about this scale of phenomena.

In other words, the larger the mass of an object, the more it becomes invisible.

This may imply a universal presence or absence of quantum states at the property level, roughly in inverse to mass, called Quantum State Theory.

A further derivation is that sub-microscopic phenomena are described by an invisibility wave.

This may in turn provide the sub-microscopic corollaries of the universal Calabi-Yau manifold.

…

The term 'coherent quantum states' was found unnecessary, as
'coherent quantum state theory' is identical in meaning to 'quantum state theory'.

FURTHER RESEARCH:

"When considering material politics, if neutrality (such as homogeneity) emerges from an exception (such as exponential efficiency), then neutrals can be exceptional. *In other words, deviation may occur invisibly. [italics added]*" – <u>General Mergence Theory</u>

THE *SCIENTIFIC THEORIES*

Dimensional Gravity Principle:

Mathematical models aside, dimensional theories might propose the following:
1. If the 3-d exists now, then the 2-d must exist in a similar way to the 1-d, and thus, in 4-d we might have 2-d gravity in a physical model. This might imply volition or some sort of causal force, or arbitration-energy.
2. Gravity affects every dimension of reality to some degree (either scalarly, or through attributes of each level).

The conclusion is that gravity equally exists in higher dimensions, *when they are physical, **and yet, they may have different properties incorporated in gravity, analogous to another dimension of gravity to the extent that gravity is important.

Latent Field Effects :

Previous knowledge touches on the effects of gravity and electromagnetism on latent fields. The most recent theories seem to be encountering a kind of Babel effect with quantum mechanics and dark energy.
Even when energy can be known conceptually, it cannot always be known empirically. And the more the empirical bounda-

ry is tested (so it appears, judging by quantum mechanics and dark energy) the less we actually know about how physics actually works. But, simultaneously, the conceptual boundary continues to progress forwards. As one physicist said, "The more we progress backwards the more we progress forwards." It would do to analyze latent fields in terms of what we don't know, since what we know is something that is frequently traversed. What we don't know is at least two things: (1) The exact condition of the latent field in the real world, and (2) The emergent property acting on this field. Therefore, it would do to define latent fields in terms of two physical properties: (1) The universal condition of nature, or else some aberration, acting as an emergent property, and (2) The best understanding of how existing properties are known to behave. From this it could be derived, given limitations on computation, that either we may derive a general rule about how latent fields behave, or we can derive the specific properties of the energy. If we take for granted that this is the type of analysis that led up to developments in quantum mechanics, then there are several options: (1) Latent fields are not a major relevant factor of analysis, (2) The specific properties of energy are exactly what is aberrant about physics, (3) Aberrant properties are created by an interaction between specific energies and the latent field, or (4) There are no aberrant properties. This may be open to some level of semantics. Thus, complexity may be an important factor in parsing the results. If what is meant by complexity is an emergent property, this gets one result. If complexity means a latent field, this gets another result. If complexity means interaction, this gets another result. If there is no aberration or latent field, this gets another result. Thus, it is important to know exactly what entities such a system is related with, and unless they are defined there is no way to know. Why wouldn't a specific sun have a specific mathematics? After all, we only have one local sun, and thus, we may have one local mathematics. If what is quantum is correlation with experience, then quantum breaks down into singularity. However, with emergent properties of entities, it is possible to theorize a multi-physics, and enact it through quantization. The answer to latent field effects appears to be that it is quantum, but that there is no defined limit on information. Information, if it is defined, is defined through singularity or the emergent properties of entities.

Physical-Chemical Dimensions:

Maybe cold is to the second dimension what heat is to the third dimension... what identity is to the first dimension. The physical patterns of the first three dimensions ... The fourth dimension thus must ignore the first two dimensions unless it involves perspective.

Cold is the paradoxical 'C'. Heat is excessive energy, which reduces to information. The fourth dimension involves paradoxical properties, the resolution of which is the expansion into many dimensions.

The Natural Philosophy of Crack-Piss:

If one follows the globules in a stream of water or urine, one notices that the central stream moves relatively further forward than many of the escaping globules. The escaping globules seem to follow, on average, a slower trajectory, first by escaping the stream, secondly by taking a longer course, and thirdly, by not being propelled physically forward by any later globules. Based on this observation, however, one can incite a theory of semantics about the laws of physics. Perhaps, for example, escaping the stream would give the globules less resistance from earlier globules. Based on this, one cannot easily simply defer to a law of common sense in describing these phenomena. For different scenarios may apply to the same case, and thus, there is no 'real' consensus about what results in the physical formulae. One solution is that all reasonable explanations take place simultaneously. But most scientists would argue that physics is less quantum than that. We don't usually argue that a bad theory creates virtual particles. Yet that is precisely what appears to be the case. One should then examine many alternative theories, more theories than have ordinarily been considered reasonable to consider, in deciding the actual properties of physical phenomena. It then follows that what is observed in the model is actually what occurs for the particles under an excited state, what I call the condition of 'crack piss'. But that is NOT to say that ideal exceptional cases (in some sense) do not exist.

Super Discrete:

In higher dimensional mathematics, a 4-dimensional or higher figure for which motion in the hyperbolic becomes less hyperbolic. In other words, the form begins to represent a lower-dimensional figure. In physics this betrays an underlying conceptual angle on physics, and could thus be used as evidence of virtual reality, or else some concept of physical complexification and decomplexification. In the second case (complex or de-complex) this could provide intriguing inroads on the relation between equational reality and physical reality, I.e. the possibility of direct representation or direct physical corollaries. In a coherent method, non-direct corollaries pose a potential problem for identification. Although they may possess opposite pairs, there may be some structural decoherence, which philosophically points towards an alienated system. It becomes necessary to adopt a special vocabulary just to convey what is meant by this form of coherence or decoherence, when otherwise the relation of entities and equations would go unstated.

Assorted Notes :

Physicists are being contradictory when they introduce the idea that an infinite universe has energy, but that in finite cases energy is *always* lost. For if energy in some cases is completely lost, then there would be no capacity to create the energy in the first place. Perhaps all forms of en-

—— THE *SCIENTIFIC THEORIES* ——

tropy are really paradoxes of energy.

The term OR can replace the term AND when relating to coherent laws which concern not just abstracta, but empirical laws. A lesson learned in The 78 Binary Laws of Physics.

Newtonian mechanics says that levers have an advantage and also that objects roll downhill, while thermodynamics says that energy disappears. But a unified principle of energy appears to say that if energy never absolutely disappears, that energy could also emerge. There is, for example, the principle of dominoes that don't have to be reset. There is no physics principle which says imperatively that dominoes have to be reset. Call it asymmetric entropy theory if you like.

If relativity implies that the amount of energy is constant then what is required for a perpetual motion machine is little other than a paradox! There would be no need to think of the entire universe as the only unit of this perpetual motion machine, because any system in which asymmetric entropy functions would have the same perpetual properties as a perpetual universe! If energy is not created somewhere, goes the principle, then if matter exists everywhere, then energy would exist no-where.

Thus, physicists are being contradictory when they introduce the idea that an infinite universe has energy, but that in finite cases energy is *always* lost. For if energy in some cases is completely lost, then there would be no capacity to create the energy in the first place. Perhaps all forms of entropy are really paradoxes of energy.

Time crystals are structures which are argued to have no observable moving parts in the third dimension, but may have dynamic properties in the fourth dimension. But they say, it is only perpetual if you can extract energy! So, why isn't a time crystal just an agglomeration of fixed properties that exist in the fourth dimension? Even if time crystals are not machines, they prove that perpetual machines are possible in the fourth dimension. Apparently, believing in time crystals is the same thing as believing perpetual motion is a reality!

Structural degeneration is solved through the pre-adoption of a structural system that is not degenerate. Where structural systems do not exist, what is implied is system building, borrowing tools, or introducing patterns other than degeneration.

——THE *SCIENTIFIC THEORIES*——

Hierarchy of Entropy:

1. Physical work degenerates.
2. Physical health degenerates.
3. Physical pleasure degenerates.
4. Principles of physics degenerate.
5. The soul of physics degenerates.
6. Physical purpose degenerates.
7. Truths of physics degenerate.
8. Physical systems degenerate.
9. Physical defenses degenerate.
10. Physical freedoms degenerate.

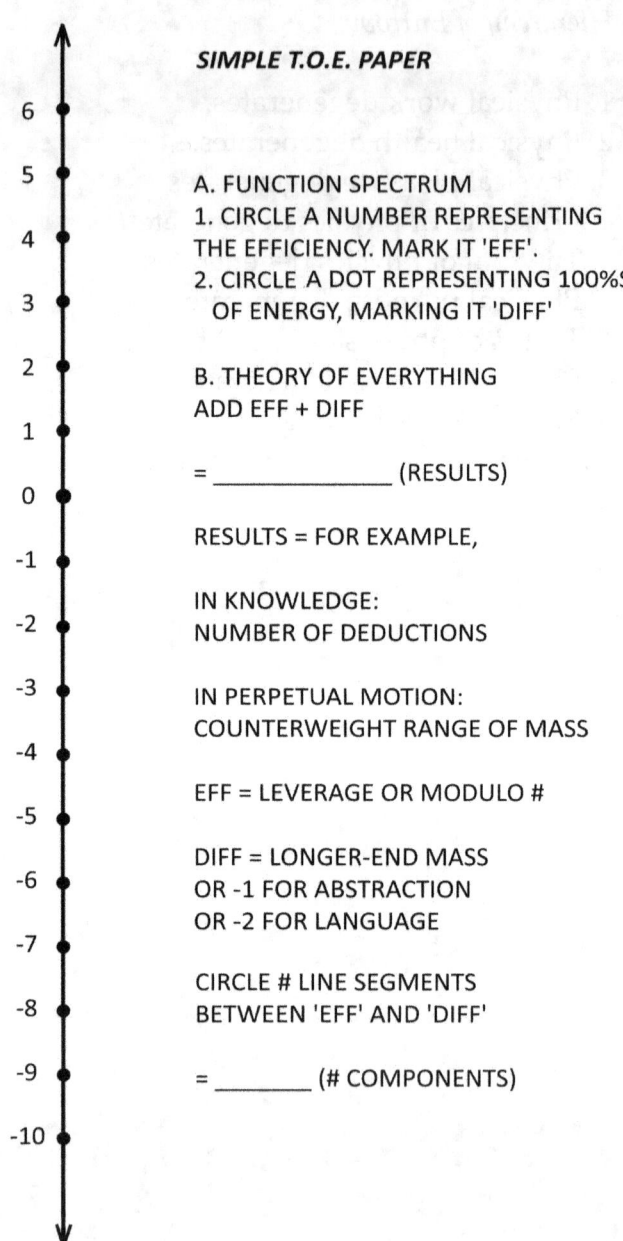

SIMPLE T.O.E. PAPER

A. FUNCTION SPECTRUM
1. CIRCLE A NUMBER REPRESENTING THE EFFICIENCY. MARK IT 'EFF'.
2. CIRCLE A DOT REPRESENTING 100%S OF ENERGY, MARKING IT 'DIFF'

B. THEORY OF EVERYTHING
ADD EFF + DIFF

= _____ (RESULTS)

RESULTS = FOR EXAMPLE,

IN KNOWLEDGE:
NUMBER OF DEDUCTIONS

IN PERPETUAL MOTION:
COUNTERWEIGHT RANGE OF MASS

EFF = LEVERAGE OR MODULO #

DIFF = LONGER-END MASS
OR -1 FOR ABSTRACTION
OR -2 FOR LANGUAGE

CIRCLE # LINE SEGMENTS
BETWEEN 'EFF' AND 'DIFF'

= _____ (# COMPONENTS)

THEORY OF EVERYTHING (TOE)

It says: "[This] Form accounts of Everything. The logic [formula] must be the most fundamental thing… So, values [sums], information [efficiency], and process [differentiation] increase in the details." --Nathan Coppedge, 2011

EQUATION

Total Results >= Total Efficiency* + Total Difference

*Where difference = results - efficiency, and where efficiency sums to < 1 if topic is acted on, and sums to > 1 if topic is acting. Since the Efficiency and Difference are correlated in the Result, the system does not commit circular reasoning if two variables are given empirically or rationally. That's the general equation with the difference being variable. In specific cases it helps to use a classification of difference see **Function Spectrum** in which the difference is fixed over different results of certain types. Remember difference = -0.5, 0, or +0.5 in the most basic cases.

COHERENT PROCESS THEORY OF EVERYTHING

MAJOR WORK 2.1: Unification

MAJOR WORK 2.2: Big Concern

MAJOR WORK 2.3: Ignore the trivial

MAJOR WORK 2.4: It might seem trivial

MAJOR WORK 2.5: Coherence

MAJOR WORK 2.6: Over-Unification

MAJOR WORK 2.7: Anti-Theoretical

MAJOR WORK 2.8: Big Explanation

MAJOR WORK 2.9: Must Reach Unity

MAJOR WORK 2.10: These are the real results

MAJOR WORK 2.11: Unity ratings

MAJOR WORK 2.12: The universe might be over-unity

MAJOR WORK 2.13: Over-Unity for TOEs

MAJOR WORK 2.14: OU must include energy

MAJOR WORK 2.15: Energy must be objective like leverage

MAJOR WORK 2.16: Efficiency + Difference

MAJOR WORK 2.17: Paroxysm

MAJOR WORK 2.18: Banded Reality

MAJOR WORK 2.19: Anti-Theory

MAJOR WORK 2.20: Really about perpetual motion computers

—The Complete Genius of Nathan Larkin Coppedge (…)

THE *SCIENTIFIC THEORIES*

NO ENERGY THEORY

December 7, 2018

[Not claiming any divine expertise here, but I don't feel obligated to say that because I do not normally assume any divine entities]

That matter [material souls] does not need to have energy! Now we know a la consciousness studies that the universe does not require energy, and therefore the information theories [of the soul] may be valid. The entire universe may not require energy. The entire universe may be magical definitions.

(—Support for Holographic Consciousness)

Bad ideas sometimes work.

Physics doesn't always work unless it includes bad ideas.

No standard is universal.

Information works, because it does not require a universal standard.

Criticism of Schrodinger Favoring Perpetual Motion by Nathan Coppedge (…)

THEORY ON THE DISAPPEARANCE OF ENERGY

Antimatter allows for perpetual motion machines. Perpetual motion may in effect create antimatter due to higher-dimensional properties. That isn't the concern of my research, however.

This may occur when energy bleeds off a closed system.

This property may be explained by matter-antimatter balance, perhaps involving hidden dimensions, perhaps related to the theory that a dimension will appear double if it manifests in a dimension two dimensions lower.

THE THEORY OF PHYSICAL ENERGY

Something to consider: gradual compound mass application. Would this create magic? Compound gravity compounding over time, distributed over time, this seems to be why matter has energy.

THE *SCIENTIFIC THEORIES*

THEORIES OF ENERGY

In certain ratios of lever acting on a slotted track, a mobile weight positioned on the track can be moved vertically and then trigger the counterweight to rise through segments where the mobile weight is unsupported. The ratios I have found are 2.4 to 3X leverage for the mobile weight (moving outward from the fulcrum) compared to 1X distance between the fulcrum and the midpoint of the counteweight, 0.5 degree approximately upwards slope of the track, and 5 - 6 degree approximately angle of the lever between the sides of the track. A large mobile weight ball can be used to activate the next lever without requiring additional gain in altitude, as permitted by the slight gain in altitude, permitting loss which can be used to trigger the next lever, as shown in 1st Fully Proven Perpetual Motion Machine.

When horizontal motion is greater than vertical, a wedge can lift a rolling object by its own weight, so long as the horizontal motion of the wedge is greater than double the vertical motion, per unit of distance.

Alternate theories: energy is not understood. Energy is created according to need. Energy is 'majority potential'. Potential energy is constantly created, and only ever *potentially* destroyed. When potential is completely destroyed, the universe ceases to exist. It is a non-state. It doesn't really cease, but ceases to be perceived. Ceases to function as an *IDEA*. But energy itself is only efficient or delegated, not absent. Energy is really 'figurative' and 'possessed' or 'depossessed' , not 'imposed' or 'spent'.

Alternate theories: Energy is a living theory. Under some conditions it has no theory. Under other conditions, it must be granted as a kind of theoretical constant. When it has no theory, it has the most potential. But when a theory is granted, there is no denying energy constants. Greater constant just means greater theory.

MATHEMATICAL CONJECTURES

PROPORTIONAL NUMBERS

This requires the invention of proportional numbers, which are in-between the finite and the infinite, and distinct from ratios. The scale of a proportional number depends on how it is used.

For example, there may be four different classifications of proportional numbers: (1) proportional number of identity are about fractions of resemblance, or inherent relations, (2) proportional numbers of concept are about systematic similarity, (3) proportional numbers of coherence are about relations between the finite and infinite, (4) proportional numbers of combination are about the relation between an operation and something variablistic.

That was just a brief sketch.

In the case of an infinite-mass universe, what is required is a proportional number between infinite mass and the scale of the universe. If the universe is infinite in any way at all, then infinite mass may be partly finite in scale, and then infinite mass can be contained by a proportion. If there is any doubt, then it sustains variables, such as the materialism of the elements of the universe, which may then have further requirements. Whether the further requirements are met may depend on exceptional interpretations.

A possible basic proof:

Possibly using proportions of infinity as an intermediate, and proving lower bounds.

I call proportions of infinity 'proportional numbers' like 1/2 or 1/4 of the X, Y plane or 1/4 or 1/8 of the X, Y, Z coordinate space.

If comparable spaces increase X infinity just like ordinals or trans-finites, this means they may be treated like mathematics even when the contents are normally assumed infinite.

THE *SCIENTIFIC THEORIES*

VARIOUS MATHEMATICAL NOTES

Infinites

1. A space the property of which is to relocate : math is then causally determined.

2. A Leier space isometric on the diagonal: proportions determine indefinite degrees in terms of coherent space.

3. A space which only appears when it is irrational in value. It appears from one perspective, but not another. Sometimes what has appeared is actually an illusion cast over a large area.

4. A form which depends on infinite bounds for its construction, for example, a set of spiraling diamonds which begins at infinity with a square shape, and at some point becomes an infinitesimally flat diamond of infinite length. Finite numbers are involved, but determining them depends on the specific configuration. What if every other pair is joined, creating a three-dimensional wedge? How does one determine the scale of spheres eaten out of these shapes? For example, if the space is isometric? If the space remains partly infinite? Etc.

5. A Klein bottle made of vanishing points. How much does it matter where the horizon lies?

6. A tube filled with rods, one for each ascending value until infinity is reached. Depending on the concept, ½ infinity might be enough to create infinity, or simply enough finite segments of infinity to reach a finite length called infinity…

Infinity has two senses:

(1) Infinity in the absolute sense, 'that cannot be surpassed,' and

(2) Infinity in a more finite sense, which CAN be surpassed.

But that means that mathematicians cannot use the word 'absolute numbers'.

Either infinity exists in the sense of not-being-surpassed, or infinity exists in a qualified sense.

If infinity exists in a qualified sense, the term 'absolute number' must be used to refer to infinity in the absolute sense of infinity.

―――THE *SCIENTIFIC THEORIES*―――

Lemma Lemma Lemma [***] -

* The first lemma appears in formal analytic logic, where it means 'some major exception' to the given variable. This is sometimes represented with the letter T, meaning you need a theory.

** The second lemma appears in advanced formal analytic logic and calculus, where it means an exception granted to the first exception, or in other words, a case in which two additional conditions hold on the initial variable or formula. This is sometimes represented by the acronym GO, which means you have to go test the equation, because you don't have as much knowledge as someone who would try more lemmas. It also means Geisweisellschaften Ordenen, or something to throw at the philosophers. A lot of algebra is involved, in other words.

*** The third lemma appears in advanced mathematics and absurd philosophy. Since the third condition can include formulas summarizing formulas interpreting a stipulation on yet another formula, going beyond this stage is arguably beyond most reasonable mathematics. It is sometimes represented by the name BAD. You have to be that good, which is not as good as some. It might also stand for 'Branded

81

Arithmetical Doctor'. Meaning you need a PhD. In Mathematics to succeed at it.

**** The fourth lemma means one thing: Einstein-ism. It may imply the potential semanticization of mathematics itself, or else the creation of a 'perfect formula'. If you're not an Einstein, the fourth lemma means something else: madness, or impossible mathematics. It is sometimes represented by the name GOOD. Which doesn't stand for anything. And you're Einstein if you know what that means. You have to be good.

***** Five lemmas is just ridiculous, and shouldn't be attempted even in a joke. It can be interpreted as meaning 'mad even to Einstein'. It is also called by the name 'MADEN' meaning crazy in German (and also standing for the five lemmas).

Pi (π) -

Pi is an irrational value reached, so far as is possible, by finding the exact circumference of a circle with a diameter of one. Thus, the circumference meets the geometric equation of "Circumference = π * Diameter" or in other words $2\pi r$.

An interesting observation is that, if a diameter of one yields π, then a circumference of one yields a diameter of $1 / \pi$. The proof is that when we input the value $1/\pi$ for the diameter in the equation "Circumference = π * Diameter" the result is $\pi * 1 / pi = 1$. The value of $1/\pi$ also fits in the diameter series $1/\pi, 1, \pi$ in which the corresponding circumferences are $1, \pi,$ and 3π. Thus, in the binary series of diameter $1/\pi$ and 1, the result is to reverse the π relation in the circumference to 1 and π.

With the area of a circle, the result is to take one half the diameter and square it in relation to π. Thus,

$(½ * 1/\pi * \pi)^2 = ¼$ is the area of a circle with a diameter of $1/\pi$.

$(½ * 1 * \pi)^2 = ¼ \pi^2$ is the area of a circle with a diameter of 1.

$(½ * \pi * \pi)^2 = ¼ \pi^4$ is the area of a circle with a diameter of π.

Proportional Numbers -

An invention lying between finite and infinite numbers. The concept assumes that one is the opposite of infinity, so as to avoid incoherence. The concepts in brackets are appropriate opposites in this system. The ratio thus expresses a literal form of division unique to proportional numbers, wherein division exists in the solution, but not in the problem. Thus, proportional numbers are a unique case in which division is derived.

For $\{\infty, 1\}$:

$\infty / 1 = \frac{1}{2}$ proportion.

$2\infty/1$ = proportional number of 2 for $(\infty, 1)$

For $\{(\infty - 1), 2\}$:

$\infty -1 / 2 = \frac{1}{2}$ proportion.

$\infty / 2$ = proportional number of $\infty - 1$ for $(\infty - 1, 2)$

For $\{(\infty - 2), 3\}$:

$(\infty - 2) / 3 = \frac{1}{2}$ proportion.

$(2\infty - 4) / 3$ = proportional number of 1 for $(\infty - 2, 3)$

For $\{(\infty - 3), 4\}$:

$(\infty - 3) / 4 = ½$ proportion.

$(2\infty - 6) / 4 =$ proportional number of 1 for $(\infty - 3, 4)$

Trans-Finite Number Primer

Part I.

There is one level between finite numbers and infinite numbers, and it is proportionality. Consider the following solid rule:

$\infty / \infty = 1$

So, $\infty / (¼ \infty) = 4$.

So, $1 / ¼$ infinity $= 4 \infty$

So, $4 = \infty * ¼ \infty$

Apparently, two infinities cannot exist together. Or, $¼ \infty = 4 \infty$. We cannot say that $¼ = 4$, so we cannot accept this as straight math, unless we incorporate some relativism.

We can, however, equally say that $\infty^2 / 4 =$

4, which is less problematic if the only constraint is not involving two infinities. That is, if the square of infinity is similar to the square root of 2, for example.

Perhaps it is more reasonable to accept the idea that they are contradictory. For example, this could be accommodated by adopting variablism: by any one value, there are no two infinities. However, when the value is not specified, the infinity is assumed to be absolute. There is no contradiction in the idea that absolute infinity contradicts itself when it is duplicated. But there is a contradiction in the idea that $¼ \infty = 4 \infty$, unless we simply think that infinity is a proportional number. However, it may be much more adequate to think of proportional numbers as yet another type of number that lies between the finites and the infinites.

Infinite Numbers are numbers countable in terms of infinites (infinite values), and constitute *numbers of reference*. In coherent mathematics, infinite values are the context for finite values, and like finite values, systemic integrity depends on the use of coherent functions. The assessment that '1' opposes infinity creates a kind of incompleteness theorem for trans-finite numbers, because it becomes difficult to represent the single unit of infinity without also creating systems. Fortunately, systems can be construed using rules of proportionality.

4∞ = coherent version of ¼ proportional number * 4

2∞ = coherent version of ½ proportional number * 2

In this sense, infinity is simply an absolute sense of numbers. This is easily applied to geometry, where we can observe that a triangle is the coherent infinite of a triangle, etc. In this sense, what a figure offers is a formal constraint, not a value. In fact, any given thing is the coherent infinite of itself, when it functions perfectly. So, now we have four main concepts to relate with: (1) Finite numbers, (2) Proportional numbers, (3) Infinite numbers, and (4) Functional numbers. Functional numbers are simply the form of ersatz vector of a concept, including all of its information. This can be treated simply by saying that a given object has the tendency to reproduce itself, or, in other words, the object contains all the information necessary for reproducing the object. However, functions become more complex if a function is complex enough to reproduce smaller constituent functions as well. In this way, math is per mutative, but also concerns the inherent properties of objects. In terms of infinites, the concern is that every object is coherent when it is a perfect function.

Indefinite Numbers are numbers with an indefinite value, such as a conditional. These may be seen as belonging to the boundary between finite and trans-finite, what may be called the logical domain. One example is finite value that depends on an infinite value for proof (a kind of nested domain, also a conditional), or a function which depends on countability above value. These may also be called in-complete numbers, and may depend on *conditional evaluation,* implicating either arbitration of sets or arbitration of functors, or else something more like statistical analysis, including graph theory (what I call the limit function of purposeful mathematics).

…

Part II.
I think I found a math system where 0/0 actually equals zero, unlike the current system, where it equals 1, and then divided by zero again, equals infinity, etc.

Mathematicians treat this as a simple conventionality, but it is a large thing to overlook, in my opinion.

In the new theory, 1 / infinity = infinity / 1 as usual (I think), since infinity = infinity

= infinity...

However, 2 infinity would equal 0, as well as all other even numbered infinities.

The result is a system in which 2 infinity / 2 infinity = 0 instead of 1.

However, all the other discrepancies appear to be overcome, with the possible exception that there is a kind of quantum flux or uncertainty to the process of counting infinities...

Logical Solutions to Mathematical Incompleteness

I have been musing about the dialogues between Einstein and Niels Bohr, and also the relations between the greatness of Einstein and the greatness of Kurt Godel. My solution to this dialogue is that there are key insights which overshadow the doubts that emerged at the time. You will excuse me if I introduce some psychology in the schematizing of the thoughts of the time, as psychology is often what is warranted in solving intractable problems. In part because mathematics and physics have often been seen as related disciplines, it can be seen that there is some factuality to the influence of hormonal competitiveness with

Einstein in the specific design of theories of his competitors, such as Godel's incompleteness.

I will leave that complex idea as an initial lemma, except that there is a particular relation between logic and mathematics that may be worth discussing. This is specifically the role between proof theory (a.k.a. Einstein), and the causal relationship between mathematics and physics.

For, we would not ordinarily say that physics is incomplete----at least I wouldn't. Instead of criticizing this point as having irrelevance out of the inherent incompleteness of physics, we can adopt a logical tool, and say that physics is semantically *complete*. For, after all, in logic we would not say that a theory is incomplete if it appeared (like Aristotle's syllogisms), to provide an explanation for any type of conclusiveness we might imagine. Nor would I say that this view is naïve, since even in science the deference is to the best available theory. We would not say science is wrong because it does not know absolutely everything. Instead, we would say that it is relatively complete. And, I argue, it is the same with physics. Unless we are being very technical about what completeness means----and I think we aren't absolutely technical here, just as the physics is not absolutely complete----then it makes sense to consider physics as though it has some degree of completeness, e.g. it is a relatively successful attempt at a complete explanation. Even if there are multiple categories of physics (relative, quantum, string theory), each category undoubtedly contributes to the completeness of physics, or the theories would be regarded as quack science.

Now that I have argued that physics is considered relatively complete, I would like to make an interesting stipulation. If physics is considered complete, and math is not considered complete, is there something that can be had here? Although math is not normally considered 'physical' ---- could it be that the immaterialism of math is groping with a primitive spiritual idea, instead of what it should be doing, which is accepting some perhaps unseen form of objectivity?

For, what is 'incompleteness' saying except that math cannot be objective? Wouldn't objective incompleteness be an oxymoronic definition? Or would it just mean that math cannot ever be complete? Then, are we saying that math cannot ever be complete in a complete sense, or are we saying that math is itself un-objective? I think no one will make the claim that math is un-objective, and asking if math can be complete in a complete sense begs the question of whether we are in fact being relativistic. For there is no sense of math apart from the objective sense, unless it becomes a math of un-objective things. But, instead of wading deeper and deeper into the sense of math as an un-objective application----which clearly leads to un-objective conclusions----the principle that truth is the obviation of the obvious gives us several options: (1) Truth is obvious, (2) Truth is about obviating, (3) Truth isn't obvious, and (4) Truth isn't about obviating. Clearly I think it is the case where math is not about obviating that seems like the weak point. But have we proved for definite that math cannot obviate truth? I think regardless of the amount of education required to learn math, it certainly can! And this goes against the principle of Gödel's incompleteness.

However, to prove for definite that there is some mathematical principle based on physics that could be foundational for math requires additional reasoning. But there is nothing which says such ideas could not be foundational in some exceptional, acceptable sense.

In my own theories on logic, I have arrived at a concept of a bounded Cartesian

Coordinate system defined by polar opposite word pairs. In this case, it was simply conceptualizing differently which permitted the context to be understandable. I would advocate a similar solution for math. The concept that some mathematical principles are complex, unavailable, or infinite may be limiting the cogency of mathematics. Furthermore, there may be some way in which infinity is not being conceptualized appropriately. My own solution has been that the trans-finite is a product of division rather than multiplication. Unless there is a concept of a whole, math will remain incomplete. But if infinity is a byproduct of multiplying and adding exponents, this assumes the consequence that there is no whole, and thus, that math is incomplete. But the process by which this occurs is not mathematical, instead, it is a more rudimentary logic that might be proven wrong, as I have shown. Certainly the concept that math is not whole does much to refute mathematics, if it comes last. But I think it could just as easily have come first. If it is a matter of cause and effect, and it could be either one, then it is clear that incompleteness is not 100% correct.

In logic, if there is something ambiguous, the process is to search for new and creative ways to solve the ambiguity. These tools are not as easy to apply in math, when the assumption is that it is a product of its products. However, math sometimes involves philosophy. It sometimes involves logic. Some of its assumptions could be wrong. And I think this is the most likely explanation for any form of absolute mathematical incompleteness.

As I mentioned, I have thought of some possible ways to support math by combining it with physics. What if, for example, there was some strength to physical arbitrariness? Once we assume that math is physical, as it may well be in some sense, we can then perhaps prove that since math is more arbitrary than physics, then math has greater support than physics! By this form of arbitrariness, what I mean is that it is not as directly influenced by causal laws. Or, more precisely, it applies to a wide range of phenomena without participating directly in their chain of causality. And, even if math *did* participate in an object's chain of causality, this would not make math less arbitrary than the objects being determined. The objects are by definition, the most determined things about observation. Math, then, whether it depends on observations, or does not depend on observations, remains less determined than matter. And, where it is less determined, so far as it does not disappear, it has a more permanent influence.

So, at this point we have a few options. Either (1) math disappears, or (2) math has some influence. But, here is the important corollary. If math has influence, math is physical. And what is physical is not incomplete. We already know, since it is less determined than matter, that it is more influential than matter. Therefore, math is more complete than matter!

Coherent Calculus

Where sums are coherent, they add up to zero relative to the origin.
They become transparent.

Coherent limit X = indefinite integral of $F(\theta)$ dx

Where $\theta = \Delta d(x, y, z)$

$\Delta d(x, y, z) + \int f[\Delta d(x, y, z)]\, dx$ = coherent calculus for 1-d.

Maybe quadratic ally / categorically

Seeing this to be somewhat relative,

$\Delta d(x, y, z)$

OR / AND

∫ f [Δ d (x, y, z)] dx =

Coherent calculus:

≈ ∫ f [Δ S] dx

S = mod dx
Parts and wholes.

≈ ∫ f [Δ mod d(x)] dx

= opp (parts and wholes) parts

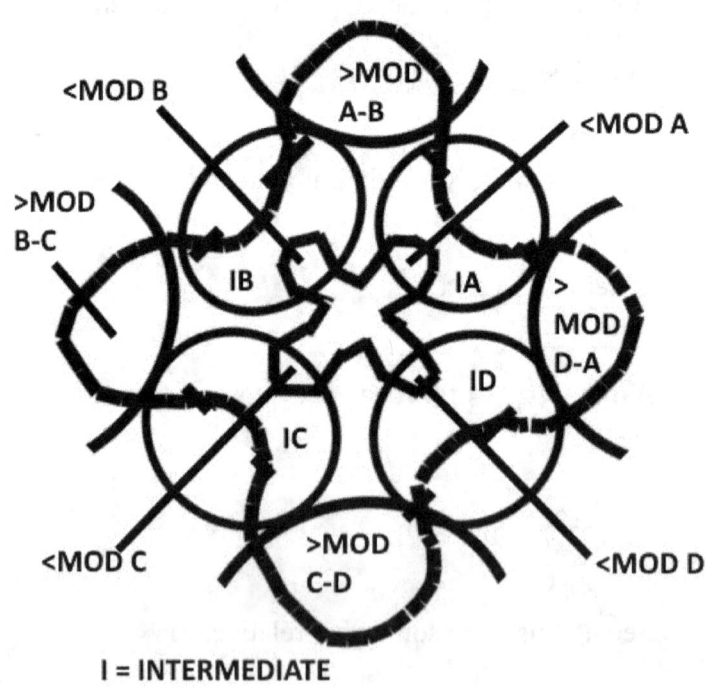

I = INTERMEDIATE

THE *SCIENTIFIC THEORIES*

(CONTINUED BELOW...)

$< \frac{1}{2}$ mod = axial whole = whole part (x, y, z) axial

$> \frac{1}{2}$ mod = part-whole opposite = +dimension

Rationicitation of intermediates.

Quasi-rationals (between coherent and variablistic).

Symbolic corollary sought.

What is done on the inside is done reverse on the outside. Outside defines how significant the inside is. If there is no inside or outside it is all f (q). Inside is the coherent +dimension. Outside is the coherent whole-part (x, y, z).

Between >mod D-A and > mod A-B is intermediate A, which is coherent A.

Between >mod A-B and >mod B-C is intermediate B, which is coherent B.

Between >mod B-C and >mod C-D is intermediate C, which is coherent C.

Between >mod C-D and > mod D-A is intermediate D, which is coherent D.

<mod A and <mod B = intermediate A - intermediate B.

<mod B and <mod C = intermediate B - intermediate C.

<mod C and <mod D = intermediate C - intermediate D.

<mod D and <mod A = intermediate D - intermediate A.

Logic operation…

Where d = dimensions,

And t is an as-yet undeclared time variable or morphism,

$|[\text{ logic } t \approx \text{correlative } \Delta \, d \ ]|$

Coherent calculus = logic t correlative Δ d

e.g. if d = 3, $\{[(b_2 - b_1) + (c_2 - c_1) + (a_2 - a_1)] - [(a_2 - a_1) + (b_2 - b_1) + (c_2 - c_1)]\} * $ logic t = coherence.

Unless the logic is dimensional, it all reduces to logic.

Depending on how you treat the matrix of numbers, other formulas may also apply. For example, perhaps the matrix involves subtraction, or perhaps the matrix involves a specific form of logical differentiation such as categories. In each of these cases, what modifies the value is either morphism or logic, and otherwise, dimensional logic.

ON SOCIAL SCIENCE

It can be observed that buildings largely consist of physical structure which defines conduction space and held space, as well as windows and other apertures, which give vantage points. In this sense, one can imagine what would be significant for the individual who stares at the wall: "A meaningful help. A job from someone else!" Employment is like an aperture in the wall, and meaning is a means of conduction between one space and the next. Finally, a theory of social architecture, and by corollary, social aesthetics! In this new sense of society it is *meaning*, formerly called information, which potentially disconnects, but more likely connects, the citizens of the metro areas.

One theory is that spirituality is a result of conformity, and intelligence is a product of bizarre outlier perceptions. But, fortunately, the bizarre outliers are common enough to create intelligence. But, at what a price! Venereal disease, misshapen bodies, not to mention neurosis and social anxiety account for a large part of the interesting perceptions that lead up to accountable intelligence. At least, they are the usual *objects* of intelligence. In this way, grotesquery accounts for a large part of the original motivation for intelligence.

―――THE *SCIENTIFIC THEORIES*―――

On the other hand, true spiritual knowledge appears to depend on *conformity*. A definition o this is 'that property that allows one to feel close enough in touch with the central stream (again, the conforming stream) of reality, that one can relax and realize the highest minded pleasures.' These pleasures are not so different from a disembodied masturbation. Indeed, they could not be alienated from the social ugliness.

The problem with virtual reality seems to amount to several things: loneliness, depravity, and an inability to command the animals that surround the 'immersed' body.

Importance is like a good germ.

Souls are needy for paradigms.

If someone lives on the 5th floor they are more likely to be happy if they think of people who live on the ground floor.

Nathan Coppedge

ON PSYCHOLOGY

If you feel normal, you will feel normal but you will never feel great. But it's actually way worse than that. You can live your life thinking it's great only to realize that it's mediocre. Or you can be the greatest person only to realize you've never been happy, or that it was all a delusion, Etc. Another argument is manic people feel 'normal' all the time.

Before every impression is fully-formed is a stage in which one may have a negative impression. It is difficult to overcome this mistake. Therefore, to overt the worst conclusions, all impressions should be positive, or concealed in sophistication or some other guise, which permits us to derive positive benefit from the material. One, the positive view, is the view of masters and children. The other, the sophisticated view, is the view of Plato's Academy. With licenses like these, history pivots on nothing more than applications.

When people learn to cope with coping, they gain happiness.

When people express what they dislike, they become aliens.

It's possible masturbation comes about through some type of asexual reproduction, like *a posteriori information propogation*.

Curious people find that insensitive food creates sensitivity, by stimulating curiosity. People who are not curious find that sensitive food promotes insensitivity, by stimulating disgust. It is senses like these that originated the concept of many-worlds, and we separate metaphysical theories from human complexity at the expense of our own understanding, if not the realizations of those forms of complexity. The reality in the sense of being separated is not differentiated, although it could easily embody unrelated things of equal complexity.

A Theory of Thought: Happiness is an exaggeration of a different feeling, which is the pinching feeling of having criteria for intelligence. Without a criteria, the happiness escapes. Without happiness, the thought is not great. A thought begins with a pinch, becomes intelligent, then becomes happy, then becomes great.

What is commonly meant by meaning defers to ultimate meaning, unless the ultimate meaning fails. Therefore, one should not defer to common meaning, unless the commonplace is itself ultimate. Therefore, there is a choice between two logics: failure and the ultimate. But what is meant by the ultimate is not just any ultimate. It begins with the good of the individual, and is elaborated through reference to systems, modes, and values.

Ultimate variation begins with the motive for change

It is possible to remember pain that was not conscious the first time: pain-as-information which threatens to emerge, just as other information threatens to submerge in fantasies.

The things that can be simulated in the mind resemble <u>exactly</u> the substances of which the mind is composed. Out of the materialism of the mind, we can suppose there are materials that are not of the mind, and yet *related* to the mind!

Indeed, the sensibility of the mind appears to be a fifth thing: beyond mind, reason, object, and its shadow.

Children benefit from hearing elementary ideas that concern a conscious learning process. Adults benefit by a variety of approaches, including stimulation & emphasis, core concepts, and unique ideas.

THE *SCIENTIFIC THEORIES*

General feelings communicate --- specific feelings are obvious.

Each person is confronting his or her own deepest problems. This explains the emotional difficulty some people have.

Passive mania is genius bottled up and waiting to be born! But when passive mania is expressed, it does not always take the form of manic-depression. There are various ways of coping with genius, just as there are various ways of coping with illness.

People who are racially or denominatively prejudiced are trying to raise their own sensory awareness. It is similar to identifying with pain. Prejudice also occurs in response to sensory cues which subject individuals associate with depersonalization or inhumanity.

Among the most suppressed theories in psychology: Aggression is correlated with excitement. Criminals may be thrill-seekers. Sexuality is correlated with victimization. Therefore, sexy people are not always functional people. Genius is correlated with madness, therefore intelligence is not strictly correlated with functionality. Stupid people (people with 90 - 100 IQ) tend to be basically functional, although often they fault their IQ in spite of this.

Sick people can still be greedy. This explains a lot about the human problem of evil. People have limited opportunities to develop the worst problems. Then most people become ill, and illness is blamed for an array of sins. People who stay well are considered blessed, and put on a plateau where they are expected to do nothing. By doing nothing, these blessed people remain blameless and good, while the remainder are sick and evil. However, for the evil people there is the excuse that illness is to blame. When good, innocent people become evil, it is declared a social ailment, a degree higher than the sickness of evil. What every moral psychologist wants to declare is that 'society was already sick', when in fact it is the psychology of evil that is sick, and physical sickness itself is no more than coincidental, or perhaps sometimes, convenient. In sum, human evil might be blamed on sickness, the inability to recover, and the worshipping of blameless idols.

The fault of science appears to be a generic psychological mistake: the idea that science is good simply because the scientific mind is good, when in reality any good mind has good results, with or without

THE *SCIENTIFIC THEORIES*

science.

People who construct their brains are the intermediate type of intelligence. The lowest type succeeds but rarely to construct anything intellectual. The highest type has moved beyond construction to the experiencing of what things really are: mathematics, wisdom, conversations, character, experience. In fact, the experiencing of the brain can lead to this form of intelligence, but requires at least the upper end of intermediate intelligence. While great intelligence first happens by fate, the determination of fate actually occurs by commitment.

On some level, the mentally weak may be attempting to make examples of themselves. There is perhaps nothing fundamentally wrong with setting an example, but without a capacity to set a real standard of what the example means, without the authority that comes with education or happiness, these examples fail to demonstrate anything that they set out to do. Perhaps ordinary people are merely successful at making examples, and this is what distinguishes them from those who are handicapped. But examples of what? What if the successful cases were examples of cruel evolutionary advantages, while the less successful examples served to illustrate dire premonitions? The strength at this point is once again the strength of example.

Suicide (when it occurs) is not just a result of bad feelings, but looking the wrong way, and then being put inside a suicidal category. This is especially true of repeat-cases. One should never want a suicidal norm, but that is what is created by this false and superficial attitude. Under this mentality, those who don't commit suicide are those who never reach that level of self-disrespect, or that level of desire for peeling away the layers of authority.

When someone thinks that someone is emotionally hurt, this could mean that things are acceptable. It could also mean that there are large questions, and it could mean that the problem is more serious than assumed. The compound of 'big question - suffering - acceptable' is often not what we think emotionally. We often assume others feel better than us, and put them on a plateau. But, in fact, they have the same feelings we have when we are not asking questions about others' emotions.

It shouldn't be fashionable to think people sound different to themselves than they do to other people.

———THE *SCIENTIFIC THEORIES*———

ON SOULS

If the mind perceives something blue, then, because the mind's essence can be nothing other than what it perceives irreducibly, then the mind must BE BLUE. But this does not mean that what is 'blue' is not open to interpretation. Thus, if blue is complex, there is nothing wrong with the mind BEING BLUE. And, if the mind perceives RED, then the mind IS ALSO RED. Thus, the properties that the minds embody are their exclusive context of experience. This is why only something that perceives what is ACTUALLY IMMORTAL, IS IMMORTAL, because all those qualities of the mind are irreducible. Similarly, if one drinks something like chocolate milk that is colored brown, one becomes partly brown. But, more likely, the color one appears to be has to do with the color of one's desire, such as who one wishes to mate with and what one DESIRES to eat or consume, since it is what one desires that MEASURES potential fulfillment. If one cannot measure a part of light, one may still be sensitive to things similar to light, and then one's skin will appear light-colored, although one will not think of it as being light, but more like something black. But, if one is sensitive instead to darkness, for example, if one is a human being with eyes, but thinks that the most important property is darkness (such as ink, paradoxes, or dark matter), then even white things may begin to seem dark-colored. One's own appearance seems to emerge as a relativity between perception and experience. But the difference between appearances and the true nature is the most striking. For one may have unacknowledged desires, natural or ultimate desires, and what one consumes is not always the extent of one's being. One can have core attributes, and core attributes are nothing other than what is perceived. In this way, the soul of the person is unalienable from perception, and it is only other souls which are capable of communicating any different idea. A soul, effectively, can only wish to change into another soul. And, the only thing that prevents improvement of life is the presence of smaller souls that have better qualities. Also, the soul can have simple and complex perceptions, and the complex perceptions often remain incommunicable. Only by elevating the mind and engaging with subtle and complex subjects, and by mixing these subjects with experiences of striking and memorable perceptions, does the mind become capable of the higher reasoning, which is the best available perception of its own intelligence.

—— THE *SCIENTIFIC THEORIES* ——

ON NATURE

It takes a perpetual motion machine to prove that time is universal primitive evolution. But, universal to what? The conventional conclusion is that time is not universal at all. But if the alternative is timelessness, then there is no alternative to some form of perpetual motion…! Otherwise, we must be 'merely' primitive, or 'merely' evolving according to the argument, or there must be some alternate hypothesis of the ultimate meaning of time. In terms of affirming the usefulness of time for humanity, I know of no better thesis, regardless of its secondary hypotheses.

Intelligence is the one thing opposing the simplicity found in survival differences. It is represented by individuation, so it is represented by the more complex character of the *image* of survival, which goes beyond behavior, and goes beyond the *soul* of behavior. Individual selection (in the sense of free-choice), thus mirrors, through intellection, the outer judgment of the species- or planetary-level idea. Symbiosis is the consciousness of similarity between self and nature, represented by survival.

What reprises the role of most organisms is their seemingly convoluted symbiosis: the way their attributes appear maximally aligned with local inhabitants. (Humans might accept volcanoes in the way some aliens accept spiky plants).

——— THE *SCIENTIFIC THEORIES* ———

ON AGING AND LONGEVITY

Forward anticipation of the eight technologies of the 2TOE25: time-travel, nirvana, unique items, Chinese, drugs, consumer items, tools, and mathematics.

- There may be a tool that is not mathematics.
- There may be a consumer item that is not a tool.
- There may be a drug that is not a consumer item.
- There may be Chinese that is not a drug.
- There may be a unique item that is not Chinese.
- There may be a nirvana that is not a unique item.
- There may be a time-travel that is not nirvana.
- There may be a mathematics that is not time-travel.

May 20, 2023: It's a new discovery: Current people may die, but immortality genes will be discovered. "My vote is for immortal genes discovery by 2049: It might be some type of change in the human nervous system, or a new type of genetically-designed organism. It may be a genetic discovery about how to generally create immortal genes."

Another process that might help immortality is to simulate rebirth and re-youth by allowing the individual to exhibit some of the properties of their mother or the early stages of growth in the human body. This may involve growing a larger body or encountering puberty more than once. It could also involve greater sexual hardiness or an increased number or orgasms. Women may live longer because of their more successful sex lives. Maybe improving the male sex life would improve male immortality. Another thing that could be done is prolonging the development of puberty, but keeping puberty equally intense at a later period.

Much like pain and pleasure, youth, it seems, under the right conditions, can be achieved at any age. It is simply a matter of the major factors that give rise to the conditionality. Youth can sometimes exist as a coincidence. I am standing by this claim. Although the incidents may be rare, it is their existence that gives rise to rumors of the fountain of youth.

Some weight gain may be necessary to live to 400 years, whereas some weight *loss* may be necessary to live to 100.

Throughout history, the best medicine for

longevity appears to have been a combination of factors, such as ultra-purified water, turmeric, whole grains, vegetables, and mental stimulants.

Strength training stops contributing to longevity when it begins to create cumbersome weight gain. Unfortunately, this is simply the point of creating stress on the body, which occurs for nearly any type of strength training. The alternative to weakness starts to look like starving the body of energy.

Efficient adaptation is the closest thing to a natural panacea for youthfulness. With the ability to quickly adapt, the body could acquire properties which are not typical of its stage of development. But, without adaptation, such characteristics depend on other factors, like brain chemistry, low stress, and nutrition. Metabolism is a major factor to target for any propensive adaptivity.

Metabolism is not just the ability to process energy, but involves a degree of physical fitness that is impossible without it. For example, metabolism may assist in eliminating varicose veins.

One major cause of late-life blindness may be chronic oxygen deprivation.

ON CONSCIOUSNESS

This is my highly original concept of consciousness, developed through my writings on philosophy and psychology. It seems plain and straightforward, but actually it is highly unconventional.

So far as I can tell, consciousness involves the senses, which are discrete organs, part of the function of which is to have an experience. They are selfish things, with no direct crossover with other organs. Consciousness is assembled from the network of organs, regardless of their actualized or concrete reality. Part of mapping experience involves following restrictive steps involving experiencing some or another part of the larger group of organs, some of which may only be modes or functions, but which nonetheless correspond, either rationally or otherwise.

This begins with mapping, or sometimes with a larger sense of the world: of the self, or other-beyond-the-self. It then proceeds inward, to concepts taken to be thoughtful or social activities, which actually involve some form of activation of the organs. The mapping of organs is essentially the metaphysics of consciousness. Even the universe is an organ.

THE *SCIENTIFIC THEORIES*

Here are the senses:

(1) The image of what has been sensed (image maps), corresponding with spotty, needy memory-actualization, like an insect.
(2) The sense of making sense (sense qua emotion and psychology), corresponding with symbols, prestige, obsession, and introspection.
(3) A sense of the sense-organs (sense qua sense), or physical reality.
(4) Sense beyond sense (larger concepts than the self), meaning spiritual and or metaphysical truth, and ultimately the best arguments.

To my knowledge, a truly comprehensive categorical concept of the senses had not been previously reached, and this form of mapping (part of which incidentally, is mapping) provides excellent criteria for the proper psychological, metaphysical, and even physical investigation of the concept of consciousness.

It's hard to imagine how pleasure would be a problem for consciousness, unless pleasure were hard or expensive to reach. But there is no sense of 'expensiveness' in nature, and difficulty usually comes with a sense of the deserved reward. Therefore, pleasure does not seem to be a problem…

How could it really be? Anyone would vote against it… It is more likely that pleasure as it has been found is a cheap alternative to true mindfulness that has been foisted as a form of blackmail, with no regard for the great losses that transpire as a consequence… Bizarre theory as it is, there is nothing which truly explains human suffereing in human terms except a failed program…

PSYCHIC EXPERIMENTS / THEORIES

If Nathan Coppedge means 'gift for stealing the edge,' and gifts involve recognition, then people are gambling that I'm immortal if I don't receive recognition immediately, e.g. because I have more than one gift, such gifting would involve continual recognition. I don't know how I would know this if I weren't psychic.

People are so obsessed with how 'Nathan' is a fool, that they don't visit my website in April, and then decide to visit more times the rest of the year. How would I know this? I must be psychic. No one has talked about it.

Magical things seem to happen at badminton, because of the idea of a 'fly,' and the graceful motions. Same goes for baseball. Applied to psychology, it requires a much higher level of behavior, such as a dinner party, psychotherapy, or telepathy.

Certain words attract birds ('thewhimsy'). Seems to have a long-term effect on the area, thus hard to prove to the scientist when no observation was recorded before the spell was cast. Since the spell begins as experimental, there is no real way to prove the magical effect is real, except to record where birds migrate to, and then ascribe magical powers to them. Apparently, the ancients already knew this.

The Dead become Christians if I throw rocks at them. If the world were fixed in its metaphorical motions, it would not be possible to create such ripples in high places. More proof of the viability of perpetual motion, because people who are easily swayed don't have strong opinions about the most important matters. Again, to know all this I must be psychic.

Bulemic people are not likely to say 'not barfing,' suggesting a kind of magical effect for certain combinations of words.

If someone is your company, they are more of an institution than they are selfless. One must differentiate between mere word-play and something else---what I think is something inherent and uncontentious.

Sometimes it's psychic pain that causes people to feel dissipated (masterful suppositional inference).

If the meme 'cram it in your face' is always produced by being manipulated by hidden information, then the victims of this (the people who get things crammed in their face) may be especially enlightened, because they tend to be antipodally different than those who are manipulated. A common example is food additives and 'loving the food'. The psychic aspect of this meme shows how dignity can be explained *sui generis*.

Nathan Coppedge

'''

—— THE *SCIENTIFIC THEORIES* ——

ALSO BY NATHAN COPPEDGE

THE DIMENSIONAL ENCYCLOPEDIA

SCIENTIFIC PAPERS [Experiments]

HISTORY OF PSYCHOLOGY

PSYCHOLOGICAL KNOWLEDGE

SYSTEMS THEORY

BOOKS WITH THE 'SPLAT' LOGO

FORTHCOMING:

SOCIAL SCIENCE KNOWLEDGE

Q.E.D.s

,,,

—NATHAN COPPEDGE —-

Nathan Coppedge or Nathan Larkin Coppedge (b.1982) formerly writing under the pennames E Terrapin and Asceticurus, is a philosopher, artist, inventor, poet, and member of the international honor society for philosophers. A prolific author with over 186 books published on Amazon, he is a perpetual motioneer, famous quotable, and internationally-selling Hyper-Cubist. A one-time member of Tesla Society UK online and PESWiki, and founder of many Facebook groups, he lives near Yale University.

www.ingramcontent.com/pod-product-compliance
Lightning Source LLC
Chambersburg PA
CBHW071820200526
45169CB00018B/473